YOUR KNOWLEDGE HAS VALUE

Bibliographic information published by the German National Library:

The German National Library lists this publication in the National Bibliography; detailed bibliographic data are available on the Internet at http://dnb.dnb.de .

Imprint:

Copyright © 2018 GRIN Verlag
Print and binding: Books on Demand GmbH, Norderstedt Germany
ISBN: 9783668958173

This book at GRIN:

https://www.grin.com/document/477201

Fatih Sahin

Development of a reverse split-ubiquitin system to characterize protein-protein interactions

GRIN Verlag

GRIN - Your knowledge has value

Since its foundation in 1998, GRIN has specialized in publishing academic texts by students, college teachers and other academics as e-book and printed book. The website www.grin.com is an ideal platform for presenting term papers, final papers, scientific essays, dissertations and specialist books.

Visit us on the internet:

http://www.grin.com/

http://www.facebook.com/grincom

http://www.twitter.com/grin_com

Johann Wolfgang Goethe-Universität Frankfurt

Development of a reverse split-ubiquitin system to characterize protein-protein interactions

Bachelor's Thesis

in Biochemistry

submitted by

Fatih Sahin

23.04.2018 – 27.08.2018

Acknowledgements

First, I would like to thank all those who have motivated and supported me during this bachelor thesis.

I would like to take this opportunity to thank Prof. Dr. Stefan Eimer, who gave me the opportunity to write my bachelor thesis with this interesting topic in his lab. His advices as well as constant interest and help accompanied me during my work.

Special thanks go to Regine Häfner, who took care of me during my work in the lab. The moral support and continuous motivation have contributed a great deal to the completion of this thesis. Thanks a lot for the patience, effort and fun I had during my time in the lab.

My thanks also go to Fenja Gawlas and Barbara Kramer for their kind help and support for other methodological as well as technical questions.

I would like to thank all members for the help with small and larger problems and of course the great atmosphere.

Most of all, I thank my family, who have supported and motivated me throughout my studies. Thank you very much for your patience!

Zusammenfassung

Ein bedeutendes Thema in der Forschung ist die Untersuchung von Wechselwirkungen zwischen Proteinen. Deshalb wurden viele Strategien entwickelt diese Interaktionen besser zu verstehen. Eine Methode ist das Hefe-zwei-Hybrid System.

Das *glo-1*(QL) Gen codiert für eine Rab-GTPase und ist in lysosomalen Organellen lokalosiert. Es wurden mehrere Effektorproteine für das glo-1 Gen gefunden, eines davon ist das C35B1.2a.

Mit Hilfe des Hefe-zwei-Hybrid Systems sollte die Interaktion zwischen GLO-1(QL) und C35B1.2a gezeigt werden. Außerdem sollte durch Promotoraustausch die Expressionsrate der Interaktionspartner dokumentiert werden.
Dieses System ist jedoch nicht für alle Proteine geeignet. Aus diesem Grund wurde das reverse Split-Ubiquitin System entwickelt, eine auf dem Hefe-zwei-Hybrid System basierte Methode um Protein-protein Interaktionen (PPI) von Membranproteinen zu untersuchen. Dabei ist das Ubiquitin in eine N-terminale Domäne (Nub) und C-terminale Domäne (Cub) gespalten wobei jede Domäne mit einem zusätzlichen Protein fusioniert ist. Die C-terminale Domäne besteht zusätzlich noch aus einem Arginin und daran anschließendes Reportergen. Kommt es zu einer Interaktion der beiden Fusionsproteine, wird diese von sogenannten Ubiquitin spezifischen Proteasen (USPs) erkannt und schneide das Ubiquitin auseinander. Anschließend wird das Reportergen aufgrund des Arginins degradiert.

In dieser Studie konnte die Interaktion nicht gezeigt werden, da keine Klone auf selektiven Platten gewachsen sind. Auch durch den Austausch der Promotoren konnte kein Wachstum festgestellt werden. Um die Interaktion zu zeigen, sollte das Split-Ubiquitin System noch weiter optimiert werden.

Abstract

An important topic in research is the study of interactions between proteins. Therefore, many strategies have been developed to better understand these interactions. One method is the yeast two-hybrid system.

The *glo-1*(QL) gene encodes a Rab GTPase and is localized in lysosomal organelles. Several effector proteins have been found for the glo-1 gene. One of them is C35B1.2a.

Using the yeast two-hybrid system, the interaction between GLO-1(QL) and C35B1.2a should be shown. In addition, the expression rate of the interaction partners should be documented by promoter exchange. However, this system is not suitable for all proteins. For this reason, the reverse split ubiquitin system was developed. A yeast-two-hybrid system-based method to study protein-protein interactions (PPI) of membrane proteins. In this case, the ubiquitin is cleaved into an N-terminal domain (Nub) and C-terminal (Cub) domain, each domain being fused to a protein. The C-terminal domain additionally consists of an arginine and subsequent reporter gene. If there is an interaction of the two fusion proteins, this is recognized by so-called ubiquitin-specific proteases (USPs) and cut the ubiquitin apart. Subsequently, the reporter gene is degraded due to the arginine.

In this study, the interaction could not be demonstrated because no clones grew on selective plates. Even by the replacement of the promoters no growth could be found. To optimize the interaction, the split ubiquitin system should be further refined.

List of Abbreviations

5-FOA	5-fluoroorotic acid
3-AT	3-amino-1,2,4-triazole
AD	activating domain
Amp	Ampiciline
BD	binding domain (also termed as DBD)
bp	basepair
DNA	desoxyribonucleic acid
E. coli	*Escherichia coli*
e.g.	*exempli gratia (for example)*
et al.	*et alii (and others)*
Fig.	Figure
GAL4	galactose-gene activating transcription factor
GEF	guanyl nucleotide exchange factor
GOI	gene of interest
His	histidine
hr	hour
IPTG	isopropyl β-D-1-thiogalactopyranoside
Kan	kanamycine
L	liter
LB	Luria-Bertani medium
Leu	leucine
M	molarity [mol/L]
Min	minute
Nub	N-terminal Ubiquitin fragment
OD	Optical density
PCR	Polymerase chain reaction
POI	Protein of interest
PPI	Protein-protein interaction
rpm	rotations per minute
SC	synthetic complete
S.cerevisiae	*Saccharomyces cerevisiae*
SSA1	Stress-Seventy subfamily A1

SSB1	Stress-Seventy subfamily B1
Tab.	Table
TEF1	translation elongation factor 1 EF1-alpha
TF	transcription factor
Trp	tryptophane
UAS	upstream activating sequence
Ub	Ubiquitin
UBPs	Ubiquitin-binding proteases
Ura	uracil
X-gal	5-bromo-4-chloro-3-indolyl-β-D-galactoside

Table of content

1. Introduction

1.1 Autophagy

Autophagy is a lysosomal catabolic pathway in eukaryotic cells for degradation and exploitation of cellular components (Ao *et al.* 2014). The process is necessary for a balance between the production of new and the degradation of old cell components. In addition, it also controls the consumption of energy sources in the depletion of cellular resources, so it plays an essential role in cell homeostasis and membrane trafficking (Awan and Deng 2014). Lysosomes are very flexible in their degradation capacities. They are single-membrane organelles that contain different hydrolytic enzymes. They can degrade small proteins up to large protein complexes (Dell'angelica et al. 2000).

1.1.1 The lysosomal pathway

The word autophagy derives from Greek and means "self-digestion". Autophagy can also be divided into chaperone-mediated autophagy, microautophagy, and macroautophagy. In the further course, only the macroautophagy will be discussed, because it represents the relevant form of autophagy for this work (Fig. 1). In the first step of macroautophagy, the substance to be degraded forms a phagophore. This material is delivered to the lysosome by a double-membrane vesicle called the autophagosome. These are fusing into an autolysosome in which the degradation takes place. Autophagy can be triggered as a kind of "stress response" or lack of nutrients or damaged organelles. Therefore, autophagy is a survival mechanism of the cell that regenerates intracellular components to regenerate ATP in the cell (Klionsky and Emr 2000, Pyo et al. 2012).

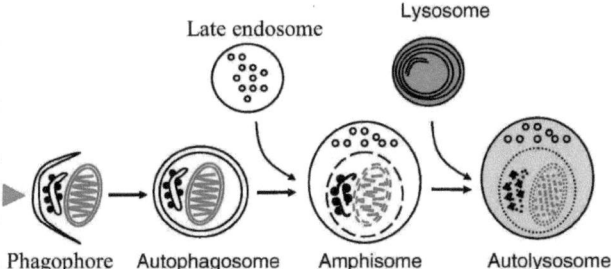

Fig. 1: The pathway of Autophagy. The material to be degraded forms a phagophore, a precursor to the actual autophagosome. The autophagosome, is forming by fusing with the late endosome an Amphisome..The formed Aamphisome fuse with the lysosome to an autolysosome. Subsequently the content will be degraded together with the autophagosome inner membrane. (edited from Eskelinen and Saftig 2009).

1. Introduction

Autophagy can be regulated by Rab GTPases. The transport vesicle traffic can be regulated, for example, by GEFs, GAPs and Rab proteins. This path is called the Rab-GEF / GAP cascade. There are specific Rab proteins that also regulate the endocytosis pathway and endosome transport.

1.2 The Rab GTPase family

The family of Rab proteins belongs to the class of Ras monomeric G-proteins. It is mostly conserved in eukaryotes and has a key role in diverse trafficking events. The eukaryotic cells have numerous membrane-enclosed organelles and compartments to handle the complex biosynthetic processes as well as endo- and exocytosis. This compartmentalization causes substances to be transported through the cell in a regulated manner. Small GTPases of the Rab family regulate such transport processes and are therefore crucial for transport specificity and organelle identity.

1.2.1 The *glo-1* gene

The *glo-1* gene, which originates from the organism of *Caenorhabditis elegans* (*C. elegans*) encodes a Rab GTPase Rab7L1, an ortholog in C.elegans, which will be analyzed. In a yeast-two-hybrid screen, several interactors of GLO-1 could be identified (R.Häfner, personal communication). One of these effectors is C35B1.2a (R.Häfner, personal communication). In prior analysis the interaction domain of C35B1.2a with GLO-1 could be narrowed down to aminoacids 95 until 301 (R.Häfner, personal communication).

1.2.2 The Rab cycle

For the transport of lipids and proteins between the different membrane-bound organelles a compartmentalization of eukaryotic cells is required. This transport is carried out by strict control of the transport vesicles. These bud from a donor compartment and then merge with an acceptor compartment. The Ras superfamily thus plays a regulatory role in vesicle budding and fusion. In regulatory GTPases the Rab proteins can have two conformations, a GTP bound state and a GDP bound state (Qin *et al.* 2017, Fig. 2). The conformation in which GTP is bound to the Rab protein is the active state. The GTP / GDP exchange factor (GEF) catalyse the nucleotide exchange. Namely, the active state is the one that interacts with effector proteins. GTPase-activating proteins (GAPs) is going to hydrolyse GTP to GDP. After the completion of these tasks, Rab GTPases are deactivated by GAPs and recycled back to the donor membrane by GDI.1. Introduction

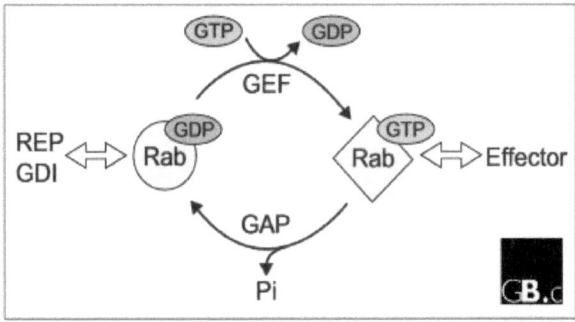

Fig. 2: The Rab GTPase cycle. Rab GTPases are present in two different conformations, a GTP- and a GDP-bound form. In order to get from the GDP state into the GTP state, the GTP / GDP exchange factor (GEF) have to catalyze the nucleotide exchange. By GTP hydrolysis, the GTP state will change to the GDP state. This reaction is catalysed by the GTPase-activating protein (GAP). The interaction with effector proteins only occurs in the GTP state. (from Harald Stenmark & Vesa M Olkkonen 2001).

1.3 Protein-protein interactions

Biochemical reactions provide a basis for a variety of protein-protein interactions and are therefore attracting much attention. An essential aspect of the elucidation as well as the regulation and function of the protein is the identification of proteins that interact with a known protein. This interest has animated the development of several biochemical and genetic approaches to identify and clone genes encoding interacting proteins. The greatest impact on interaction cloning methodology appears in the development of the Yeast two-hybrid system. ("The Yeast Two-Hybrid System", 1st edition, Paul L. Bartel and Stanley Fields, 1997)

1.4 Yeast Two Hybrid System

The Yeast Two Hybrid System is a technique used to study protein-protein interactions in yeast, especially in *Saccharomyces Cerevisiae (S. Cerevisiae)*, discovered by Fields & Song in 1989 (Fields and Song 1989) The method also can be used to define or test the domains, which are necessary for the interaction of two proteins. The peculiarity of this system is the activation by proteins that initiate transcription (Chang Bai et al. 1996). The idea is based on a transcription factor called GAL4. This transcription factor consists of a DNA binding domain (BD) and a DNA activation domain (AD) (Ma and Ptashne 1985). These two domains can also be separated without losing their function. The BD binds certain sequences so called Upstream activating sequences (UAS). This activation complex allows the attachment of RNA polymerase and is therefore essential for the start of transcription (Brent & Ptashne 1985).

1. Introduction

For the investigation of protein-protein interactions, the properties of the GAL4 fragments were exploited. The activation domain does not have to adhere directly to the DNA binding domain but must be close to the activator sequence. In this way, the RNA polymerase can attach to the DNA. Therefore, it is possible to fuse additional proteins to the single domain. The fusion protein, which consists of the DNA binding domain and a protein X, is referred to as bait protein. The fusion protein, which consists of the DNA activation domain and a protein Y, is called Prey protein (Fields and Song 1989). If the interaction between Bait and Prey occurs, the activation complex forms and the RNA polymerase can bind to the DNA. The transcription of the reporter genes is initiated. (Brent and Ptashne 1985) GAL4 and GAL80 are regulatory proteins. The GAL4 protein activates the transcription by binding the DNA binding domain. The GAL80 protein inhibits them by binding the DNA activation domain (Guthrie & Fink 1991, Fig. 3).

The lac operon with its GAL4 activator sequence is used for this system. Reporter genes are regulated via the operon, in the present system these are lacZ for the blue-white screening, and HIS3 as auxotrophic markers (Ma and Ptashne, 1987). Blue-white screening is a widely used and popular method among reporter genes. It is based on the lac operon, more precisely on the lacZ gene of the operon. The gene codes for β-galactosidase, an enzyme which hydrolyses lactose into galactose and glucose (Stagljar et al. 1998).

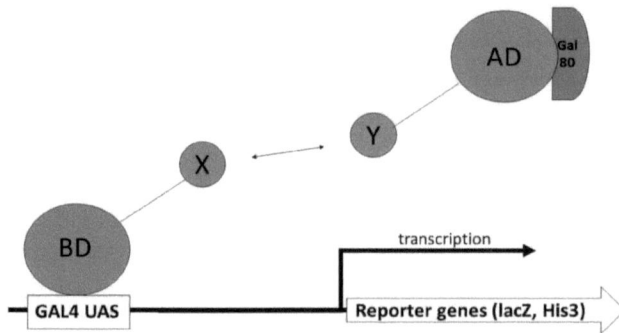

Fig. 3: Principle of the Yeast two hybrid system. As bait protein a protein X is fused to the DNA-binding domain (BD). The transcriptional DNA-activation domain is fused to the protein Y to function as a prey protein. The bait protein binds to the GAL4 upstream activating sequence (UAS). The interaction between the bait and prey protein leads to the recruitment of the RNA polymerase and the transcription of the repoter genes. The GAL4 protein activates the transcription and the GAL80 protein inhibits the transcription.

1. Introduction

1.4.1 The Split Ubiquitin System

With the split-ubiquitin system, designed by Johnsson and Varshavski in 1994, protein-protein interactions between membrane proteins can be characterized (Johnsson and Varshavski 1994). In contrast to the classical yeast two-hybrid system, the interaction does not have to take place in the nucleus but can be studied directly on the cellular membrane.

An important function of ubiquitin is the control of defective or damaged proteins. These proteins are labeled with a poly-ubiquitin tail and transported to the proteasome. There will be recognized by certain ubiquitin-specific proteases (USPs), which ultimately cut and inactivate the protein.

To use the split ubiquitin system in analysing protein-protein interactions ubiquitin is divided into two independent domains, a N-terminal (Nub) and a C-terminal domain (Cub). These two domains have a basic affinity for each other, so they can spontaneously reassociate to native ubiquitin. Mutations were placed into Nub by replacing the aminoacids isoleucine into glycine, alanine or valine. This will reduce the affinity for the C-terminal Ubiquitin and thereby suppress the spontaneous reassembly of Cub and mutated Nub (Johnsson & Varshavsky 1994). A reassociation is only observed, if the corresponding parts are near each other by the fusion two two interacting proteins P1 and P2 (Fig. 4).

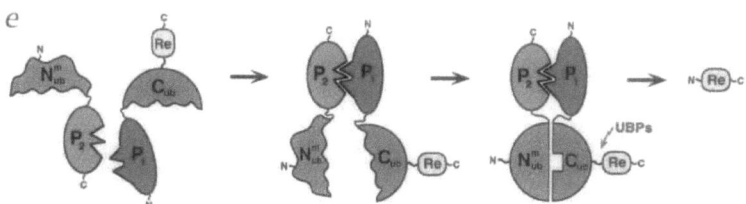

Fig. 4: The split ubiquitin system: The mutated N-terminal ubiquitin is fused to protein P2 and the C-terminal ubiquitin is fused to the protein P1. If an interaction between P1 and P2 occurs, the ubiquitin specific proteases (USP) will recognize that and cleave the reporter gene and makes it inactive. (from Johnsson & Varshavsky 1994).

Subsequently, a Cub-attached reporter is cleaved off from split-ubiquitin. There are different ways in which the split ubiquitin system can be applied. An example is the rUra3p based split ubiquitin system.

1.4.2 The rUra3p based split-ubiquitin system

In the R-Ura3p based split-ubiquitin system, a Ura3p protein is fused to the Cub. This Protein is also attached to the bait protein. If the bait and prey protein interact to each other, these Ubiquitin specific proteases recognize this interaction (Dirnberger et al. 2008). There is a glycine-glycine motif on the sequence between the C-terminal ubiquitin and the R-Ura3p protein, which is recognized and subsequently cleaved by the proteases, a so-called glycine-glycine double motif (Hershko, 2005). After cleavage a destabilized protein with a liberated N-terminal arginine on R-Ura3p Protein is found. This protein is quickly degraded by the N-end rule. This rule is essential for cellular processes and plays a role in removal or degradation of misfolded proteins. The R-Ura3p has no function anymore. If the prey and bait protein are not interacting, the R-Ura3p is fully functional and converts the 5-Fluoroorotic acid (5-FOA) into a toxic metabolite. In this way, the cells are not able to grow on plates supplemented with 5-FOA (Dirnberger et al. 2008, Fig. 5). The 3-amino-1,2,4-triazole (3-AT) measures protein interaction. The interaction only can be detected by activation of the HIS3 reporter gene. To ultimately detect the interaction, the proteins must interact on a medium lacking this amino acid. The 3-AT is a histidine analogue, which is added to the medium. This allows you to set the stringency of the reporter. The 3-AT acts as a competitive inhibitor of the *HIS3* reporter gene by competing in the active site of the enzyme.

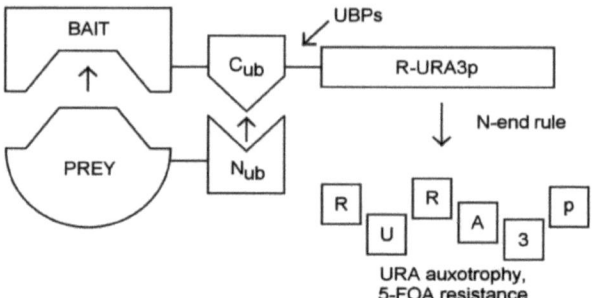

Fig. 5: R-URA3p-based split ubiquitin system. If the bait protein interacts with the prey protein, the ubiquitin specific proteases will degrade the *R-URA3p* reporter gene. This makes the yeast cell resistant to 5-Fluoroorotic acid (5-FOA) and accordingly URA-auxotroph (from Dirnberger *et al.,* 2008)

1. Introduction

1.4.3 Screening with the reverse split-ubiquitin system

Fig. 6 shows the construct order of vectors containing a C-terminal ubiquitin. This should interact with the N-terminal ubiquitin. The first vector contains the translational elongation factor EF1-alpha (TEF1) promoter in the first place. In second place is the gene of interest (GOI), followed by the C-terminal ubiquitin. There is a glycine double motif with arginine and attached *HIS3* reporter gene. The second vector and third vector contain the same genes except that the TEF1 promoter was replaced by the chaperonin promoter SSA1 and SSB1.

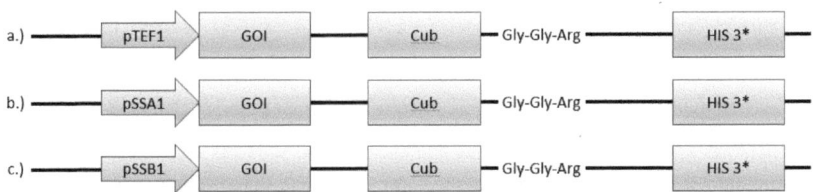

Fig. 6: Desired pDB Trp plasmids. The genes are in the following order in the plasmid: a) translational elongation factor EF1-alpha (TEF1) promotor, Gene of interest (GOI), C-terminal Ubiquitin, glycine doublemotif and HIS3 reporter gene b) Chaperonin promotor SSA1, Gene of interest (GOI), C-terminal Ubiquitin, glycine doublemotif and HIS3 reporter gene   c) Chaperonin promotor SSB1, Gene of interest (GOI), C-terminal Ubiquitin, glycine doublemotif and HIS3 reporter gene

The Cub will interact with Nub. If an Interaction occurs, the USP recognize the double motif glycine-glycine and release the R-HIS3p, which will be degraded by the N-end rule. If there is no interaction, the USP does not recognize any motifs and the *HIS3* reporter gene will be translated.

1.5 Aim

The aim of this work was to establish a variant of the split-ubiquitin system. In addition, the system should be further optimized by a promoter exchange to confirm the interaction between GLO-1 and C35B1.2a.

2. Material and Methods:

2.1 Yeast strains:

The following yeast strains were used for the split ubiquitin system.

Tab. 1: Yeast strains for the transformation. Shown are the yeast strains and its genotype with which the yeast transformation was performed.

Yeast strains	Genotype
AH109 Clontech	MATa, trp1-901, leu2-3, 112, ura3-52, his3-200, gal4Δ, gal80Δ, LYS2::GAL1UAS-GAL1TATA-HIS3, GAL2UAS-GAL2TATA-ADE2, URA3::MEL1UAS-MEL1 TATA-lacZ
BY4741[b]	his3Δ1; leu2Δ0; met15Δ0; ura3Δ0

2.2 Yeast transformation

YAPD medium was used as a growth medium for the strains *AH109* Clontech and *BY4741[b]*. This was shaken overnight at 30°C at 200 rpm until the stationary phase. Subsequently, the medium was diluted 1: 5 in fresh YAPD medium and stirred for a further 3 hours (hr) at 130 rpm. Meanwhile, 2 mg/μl salmon sperm (Invitrogen, #15632-011) DNA was boiled in the heating block at 99°C for 15 minutes and put immediately on ice. A DNA Mix was prepared. For this, 800 ng of each plasmid were needed for the transformation and used in a final volume of 34μl. To enhance the transformation efficiency 240μl PEG 3350 (50%) and 36μl lithium acetate (1M) were added.

The Yeast was centrifuged shortly at 8000 rpm. 50 μl of the boiled salmon sperm DNA was resuspended with the DNA Mix. The Yeast was centrifuged at 8000 rpm again and the cell pellet was overlayed with the DNA Mix containing the boiled salmon sperm DNA and vortexted shortly. The Mix was incubated at 42 °C for 1:30 hr at 500 rpm.

The cells were centrifuged at 800 rpm for 30 seconds. The pellet was resuspended very carefully with 800 μl H_2O. 400 μl of the cells were applied to SC plates containing no leucine and tryptophan. (See App. SC Medium). Glass beads were used to cover the cells on the plates. The plates were incubated for at least 2 days at 30 °C.

2.3 Polymerase Chain reaction:

For the Polymerase Chain Reaction (PCR) KAPA High Fidelity (KK2601, Biosystems) and Phusion high-fidelity DNA polymerase (F530S, ThermoFischer Scientific), Polymerase, were used as described by the manufacturer. The primers used are shown in Appendix, Tab. 6.

2. Material and Methods:

Tab. 2: PCR approach with KAPA High Fidelity DNA Polymerase.

6 µl KAPA High Fidelity DNA Polymerase
0,4 µl forward Primer (10 µM)
0,4 µl reverse Primer (10 µM)
1 µl genomic DNA
4,2 µl H₂O

Tab. 3: PCR cycle parameter for KAPA High Fidelity DNA Polymerase

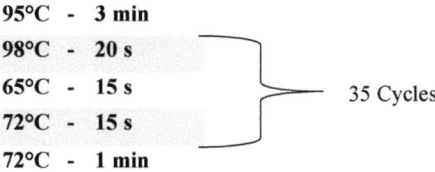

95°C - 3 min
98°C - 20 s
65°C - 15 s   35 Cycles
72°C - 15 s
72°C - 1 min

Tab. 4: PCR approach with Phusion high-fidelity DNA polymerase.

10 µl 5x PCR buffer GC
1 µl dNTP (10 mM)
2,5 µl forward Primer (10 µM)
2,5 µl reverse Primer (10 µM)
0,5 µl genomic DNA
0,5 µl Phusion high-fidelity DNA polymerase
33 µl H₂O

Tab 5: PCR cycle parameter for Phusion high-fidelity DNA polymerase.

98°C - 30 s
98°C - 10 s
53°C - 30 s
72°C - 10 min
10°C - ∞

2. Material and Methods:

2.4 Ligation into the pGEMT vector

The pGEMT vector system (Promega, Mannheim) is very suitable for cloning PCR amplificates into a vector. To check whether the PCR product was incorporated into the vector, the blue white selection was used.

2.5 Cloning into target vectors:

For the preparative digestion always a 20 µl batch was used. This included 2-3 µg DNA template, 2 µl 10x buffer and 0.5 µl enzymes (10 µg / µl) and H_2O.

After addition, the digest was incubated for 2 hours at 37°C and applied to a 1% agarosegel electrophoresis. The bands were cut out with a scalpel under UV light and extracted according to instructions from the manufacturer (GenElute ™ Gel Extraction Kit, Sigma Aldrich).

2.6 Bacterial Transformation

Table 5: Used bacterial strain with its genotype.

Bacterial strain	genotype
E. coli DH5a	F⁻ endA1 glnV44 thi-1 recA1 relA1 gyrA96 deoR nupG purB20 φ80dlacZΔM15 Δ(lacZYA-argF)U169, hsdR17($r_K^- m_K^+$), λ⁻

For the transformation of the ligation mixtures by heat shock competent *DH5a* cells were used, which were previously thawed on ice. The ligation mixture was added to the cells and incubated for 30 minutes on ice. They were exposed to a heat shock of 42°C for 45-50 seconds and placed on ice for 2 minutes. Subsequently, liquid LB medium without antibiotic was added and shaken at 37°C for 1,5 hr if necessary. The plating was carried out on LB selection plates, which were incubated overnight at 37°C.

2.7 DNA Mini-Preparation

Overnight cultures are scheduled the day before the mini preparation. Each of them contains a bacterial colony in 5 ml LB medium with the appropriate antibiotic and are then stirred at 37°C overnight in the incubator at 200 rpm. 4 ml of this overnight culture were harvested the next day by centrifugation at 6000 rpm for 3 minutes in an Eppendorf Centrifuge. The mini preparations were made according to instructions from the manufacturer PureLink™ Quick Plasmid Miniprep Kit (Invitrogen). To get a better amount of DNA, sequencing verified plasmids were purified according to the instructions of GeneJET Plasmid Miniprep Kit (#K0692, ThermoFisher Scientific™). Concentrations were measured with the NanoVue™

2. Material and Methods:

Plus Spectrophotometer (GE Healthcare). All constructs were screen in control digestion (see Fig. 8,10,12 and 14) and sequenced before further usage.

2.8 Control digestion

To screen the minipreparations, a test digestion was performed. For this, 500 ng DNA from each Minipreparation were cut with selected enzymes and analyzed in agarose gel electrophoresis.

2.9 Sequencing

To check the cloned vectors for correctness, they were sequenced according to the "Sanger" method at Eurofins Genomics in Ebersberg. The samples were prepared according to the manufacturer's instructions. The nucleotides used for this are listed in the Appendix in Table 2.

3. Results

3.1 Creating vectors with TEF1 promoter

3.1.1 GLO-1 and C35B1.2a

To develop a special Yeast Two Hybrid System, the "bait" and "prey" genes were first cloned into the appropriate vectors containing a TEF1 promoter. *glo-1*(QL) was used as prey protein and was cloned into a pNub frame 2 vector which containes the N-terminal Ubiquitin. The *glo-1*(QL) form is a dominant active form. Over time, several effector proteins interacted with the protein. One of them is C35b1.2a, which was cloned into a pDB Trp vector and act as bait protein (R.Häfner, personal communication). This vector also contains the C-terminal Ubiquitin (Fig.7).

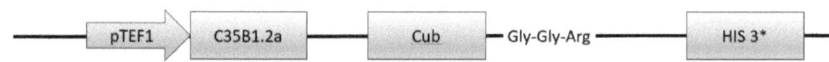

Fig. 8: pDB Trp vector with a TEF1 promotor and C35B1.2a as the gene of interest. It also contains the C-terminal Ubiquitin, followed by a glycine-glycine-arginine sequence and the reporter gene His3.

Subsequently, a test digestion was practiced checking whether the gene was completely cloned into the vector. In the test digestion for bait, restriction enzyme MluI was used. Here two bands were expected at 7631 bp and 730 bp for positive clones. For prey, three bands were expected at 3887 bp, 3484 bp and at 1274 bp. (Fig. 8A). For pNub frame 2 – glo-1(QL), the gel showed the expected bands. It also showed one other band at under 2000 bp. It was then sent away for sequencing. In the test digestion of the pDBTrp-TEF1-C35B1.2a, restriction enzyme MluI was used. Here two bands were expected at 7631 bp and 730 bp. Again, the gel showed the expected bands (Fig. 8B).

3. Results

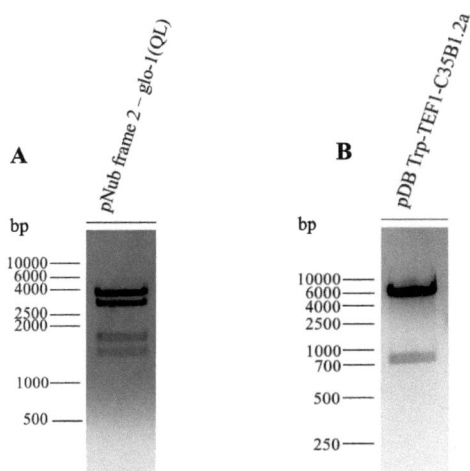

Fig. 9: Test digestion of pNub frame 2- glo-1(QL) and pDBTrp TEF1 – C35B1.2a applied on a gel electrophoresis. (A) Test digestion with EcoRV. Three Bands were expected at positions 3887 bp, 3484 bp and at 1274 bp. The gel showed the three expected bands for correct insertion. It also showed one other band at under 2000 bp. (B) Test digestion with MluI. For correct insertion two bands were expected at positions 7631 bp and 730 bp. The gel showed the two expected bands.

3.1.2 C35B1.2a (Aa95-301)

In previous analysis the interaction domain of C35B1.2a to GLO-1(QL) could be narrowed down to Aminoacids 95-301 (R. Häfner, personal communication). We also wanted to analyse this Interaction in the Reverse Split-Ubiquitin System. For this, it was cloned into the pDB Trp vector with TEF1 promotor (Fig. 9). For test digestion restriction enzyme Bsp119I was used. For positive clones, two bands at 6412 bp and 892 bp are expected. The gel showed two bands (Fig.10).

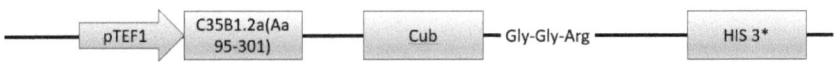

Fig. 9 Created pDB Trp vector with a TEF1 promotor and C35B1.2a (Aa 95-301) as the gene of interest. It also contains the C-terminal Ubiquitin, followed by a glycine-glycine-arginine sequence and the reporter gene His3.

3. Results

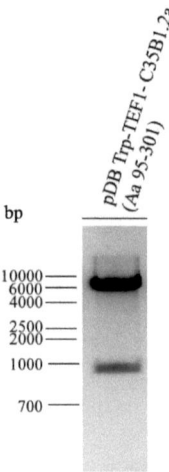

Fig. 10: Test digestion of pDB Trp - TEF1 – C35B1.2a (Aa 95-301) applied on a gel electrophoresis. For the digest the enzyme Bsp119I was used. Two Bands were expected at positions 6412 bp and 892 bp. The gel showed the two expected bands for correct insertion.

3.2 Creating vectors with SSA1 and SSB1 promoter

As it might be that the TEF1 promoter is leading to a strong expression (Th. Blume, 2016), we decided to express C35B1.2a also with weaker promoters, Stress-Seventy subfamily A1 (SSA1) and Stress-Seventy subfamily B1 (SSB1) (Fig.11).

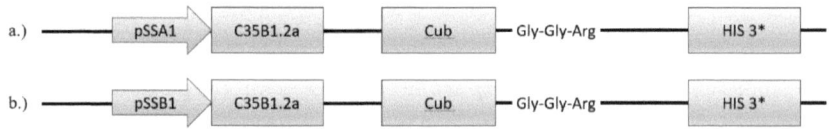

Fig. 11: pDB Trp vector with C35B1.2a as the gene of interest. It contains the C-terminal Ubiquitin, followed by a glycine-glycine-arginine sequence and the reporter gene His3 with a.) SSA1 promoter b.) SSB1 promotor

The promoters SSA1 and SSB1 were cloned into the pDB Trp - TEF1 - C35B1.2a Cub His3* Vector. For the SSA1 promoter, the enzyme CaiI was used for the test digestion and two bands were expected at 7626 bp and at 946 bp. The gel showed the expected two bands (Fig. 12A). For the SSB1 promoter, the enzyme Bsp119I was used in the test digestion. Upon successful cloning, 3 bands were obtained at 5758 bp, 1675 bp and 997 bp. The gel showed the three bands (Fig. 12B).

3. Results

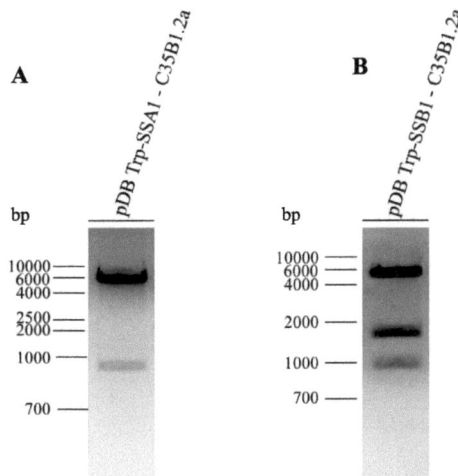

Fig. 12: Test digestion of pDB Trp SSA1 – C35B1.2a and pDB Trp SSB1 – C35B1.2a applied on a gel electrophoresis.
(A) Test digestion with CaiI. Two Bands were expected at positions 7626 bp and at 946 bp. The gel showed the two expected bands for correct insertion. B) Test digestion with Bsp119I. For correct insertion three bands were expected at positions 5758 bp, 1675 bp and 997 bp. Here the gel showed the expected three bands at their corresponding positions.

In addition, to exchange the TEF1 promotor with SSA1 and SSB1, there were also cloned into the bait vector pDB Trp - TEF1 - C35B1.2a (Aa-95-301) Cub His3 *.

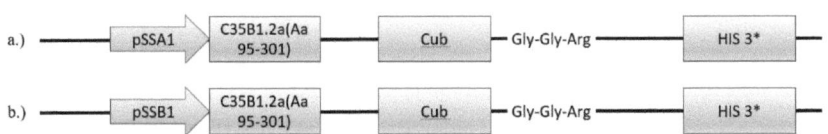

Fig. 13: pDB Trp vector with C35B1.2a (Aa 95-301) as the gene of interest. It contains the C-terminal Ubiquitin, followed by a glycine-glycine-arginine sequence and the reporter gene His3 with a.) SSA1 promoter b.) SSB1 promotor

The enzyme AlwNI was used for the test digestion and 2 bands were expected at successful cloning of the SSA1 promoter at the positions 6569 bp and 946 bp. Here, the gel shows four bands instead of two. One band is just below 10000 bp and the other below 2000 bp. (Fig. 14A). When the SSB1 promoter was cloned into the vector, the enzyme Bsp119I was used for test digestion and three bands were expected for a successful cloning insert at positions 5758 bp,

3. Results

892 bp and 728 bp. Again, the expected result with the three bands were shown in the gel (Fig. 14B).

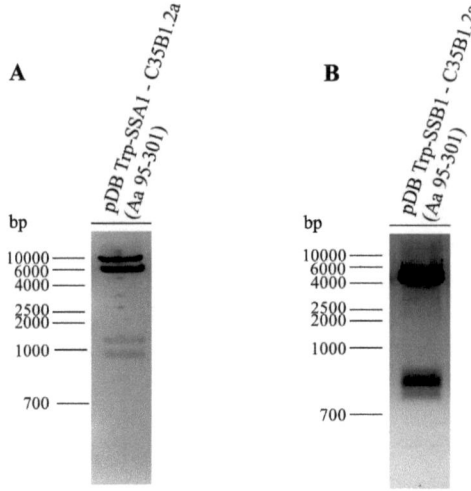

Fig. 14: Test digestion of pDB Trp - SSA1 – C35B1.2a (Aas 95-301) and pDB Trp - SSB1 – C35B1.2a (Aa95-301) applied on a gel electrophoresis. (A) Test digestion with AlwNI. Two Bands were expected at positions 6569 bp and 946 bp. The gel showed the two expected bands as duplicates for correct insertion. (B) Test digestion with Bsp119I. For correct insertion three bands were expected at positions 5758 bp, 892 bp and 728 bp. Here the gel showed the expected three bands at their corresponding positions.

3.3 Yeast Two Hybrid Assay

To test the interaction between GLO-1(QL) and C35B1.2a as well as GLO-1(QL) and C35B1.2a (Aa95-301), the vectors with the respective interaction partners were transformed into the yeast strain *BY4741*. Thereafter, it was applied to special yeast plates lacking the amino acids leucine, tryptophan and histidine. The 3-AT is a competitive inhibitor for the His3. Before spotting on the yeast plates, the OD_{600} was set to 0.5.

The (SC) -Leu-Trp plate served only as a control for the inclusion of the plasmids containing Leucin and Tryptophane. All selected clones should grow on these plates. The positive controls used were vectors that were extensively analyzed and tested for interaction in previous experiments (Th. Blume, 2016). The negative controls were partners that were designed so they could not interact with each other. The clones containing the bait and prey constructs, GLO-1(QL) and C35B1.2a, should show increasing growth on the yeast plates with higher 3-AT concentrations.

3. Results

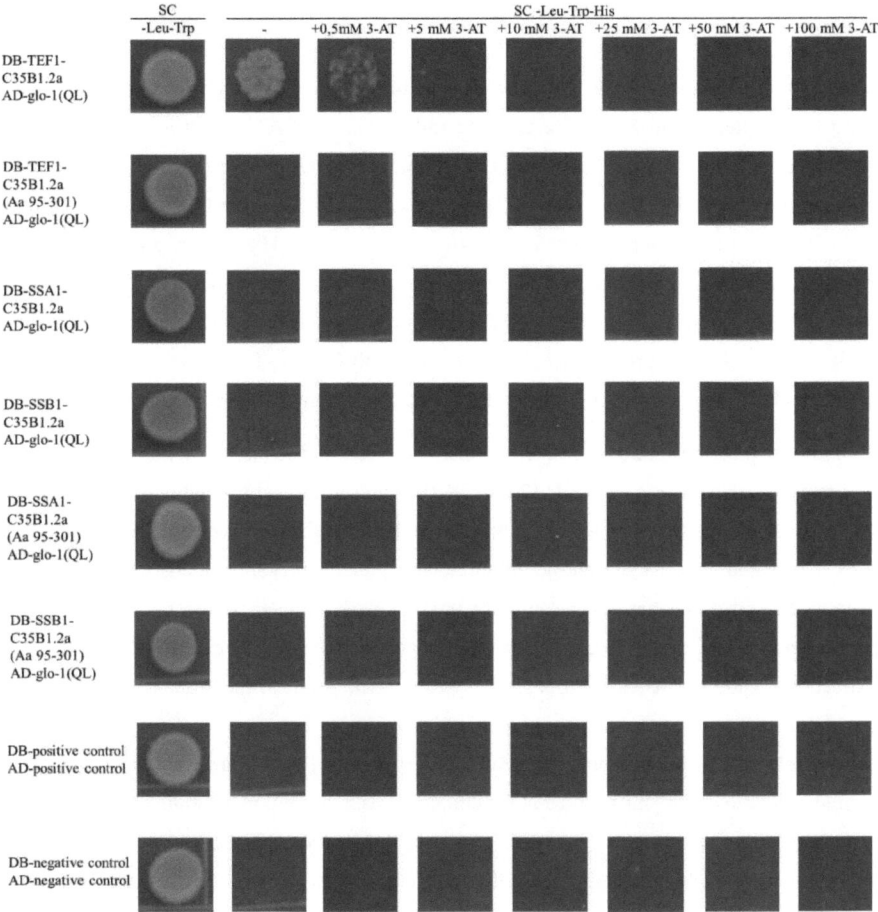

Fig. 15: Spotted yeast strain with bait and prey construct. The transformed yeast strains were spotted on SC+His, SC-His and SC-His + 3-AT plates. The 3-AT concentration ranges between 0,5 mM and 100 mM. All clones showed growth in SC+His. TEF1 – C35B1.2a and glo-1(QL) showed weak growth in SC-His + 0,5 mM 3-AT and SC-His + 5 mM 3-AT. All other clones showed no growth on SC-His + 3-AT plates.

The Yeast plates were spotted with a 3-AT concentration of 0.5 mM, 5 mM, 10 mM, 25 mM, 50 mM and 100 mM. The results showed growth on -Leu-Trp plates everywhere (Fig. 15). The interaction partners TEF1 - C35B1.2a and glo-1(QL) showed weak growth up to a 3-AT concentration of 0.5mM. There was no growth in TEF1 - C35B1.2a (Aa 95-301) with glo-1(QL). Likewise, the interaction partners with the replaced promoters SSA1 and SSB1 as well as the positive and negative controls showed no growth at all.

4. Discussion

In this thesis a new variant of the Split-Ubiqiquitin system was established to investigate protein-protein interactions. For this, the proteins to be interacted were transformed into the yeast and spotted on plates lacking histidine and containing 3-AT, which inhibits the *HIS3* reporter gene competitive. In addition, the split-ubiquitin system was used to test several effector proteins for the GLO-1 protein. The glo-1 gene codes for a Rab GTPase in *C. elegans* and is localized to lysosomal related organelles (LROs). Rab GTPases play a central role in vesicular transport and the trans golgi network (Hermann *et al.* 2005). The *glo-1*(QL) form is a mutant that is dominant active. One of these effector proteins is the C35B1.2a protein and the C35B1.2a (Aa 95-301) protein that should interact with GLO-1. The system should also be optimized by exchange the TEF1 promoter with the chaperonin promoters SSA1 and SSB1.

To allow the proteins to interact with each other, they were cloned into the appropriate vectors. The test digestions in Fig. 8B, 10, 12 and 14 (B) showed the corresponding bands in each cloning. This means, that each insert has been properly inserted into the vector. In Fig. 8A, there were performed four bands instead of three. It could be probably undigested inserts or contaminants, that may have been caused by unclean work or pipetting errors. In addition, the restriction enzymes might be contaminated or something wrong was excised during the gel extraction (or the theoretical card is simply wrong). The result of the sequencing, however, showed no mutation. Also, in Fig. 14A two bands were to be expected. The gel, however, showed four. This is because on an already dried gel, another liquid agarose gel was accidently poured into a chamber. This resulted in the double bands in the gel. In addition, the sequencing also showed no frameshifts. To investigate Proteins that interact with each other, they must be spotted on plates lacking histidine. The USP recognize this interaction from the both Ubiquitin domains and cut the sequence on the C-terminal glycine-glycine motif. The now free-standing R-His3p fragment is then degraded by N-degrons following the N-end rule (Bachmair *et al.* 1986, Varshavsky 1995).

In Fig. 15 it could be seen that growth was everywhere on the SC+His plates. This suggests, that the transformation of the cloned plasmids into the yeast was successful. Since histidine is included in the plates, the clones can still grow even if the reporter gene is degraded. In addition, the figure showed that the clones showed TEF1 - C35B1.2a with GLO-1 weakened growth. There should not be any growth here. An explanation for this would be, for example, that the activity of the TEF1 promoter is very strong (Funk et al, 1995, Kitamoto *et al.* 1998). There is an overexpression of the C-terminal ubiquitin half. As a result, the translation of the His3p is strongly promoted and can no longer harmonize with the N-terminal ubiquitin half. Therefore, the reporter gene cannot be completely degraded by the N-end rule and ensures that, despite an interaction, a growth on the plate with 0.5 mM 3-AT can be seen.

To avoid this problem, the TEF1 promoter was next exchanged and replaced by the weaker chaperonin promoters SSA1, and SSB1 for an efficient expression (Peng *et al.* 2015). The yeasts can only grow on the plate with the addition of glucose. The SSB1 promoter has less activity compared to the SSA1 promoter. Unfortunately, here as well as the controls, there was no growth on any plate. This could be, because the interaction of bait and prey protein did not work. One reason for this could be, that while spotting on the yeast plates something went wrong. It could also be due to the yeast plates themselves. Another possibility is, that something went wrong while pouring the plates. Other reasons are, that the promoters SSA1, SSB1 and TEF1 in the controls may have been inactivated for some reason. The system can be further optimized by replacing the promoter again. This can be the copper-dependent CUP1 promoter (Peng *et al.* 2015, Koller et al. 2000). The SSA1 promoter indicates increased activity when grown on ethanol. So, one possibility is to add ethanol and see if there's an interaction (Peng *et al.* 2015). The promoter in the pNub frame 2 vector can also be replaced by a stronger one, like the TEF1 Promoter. Another possibility would be the r-Ura3p based split ubiquitin system. In this case, the r-Ura3p is coupled to the C-terminal ubiquitin. By adding 5-FOA, which is converted into a toxic metabolite, clones can be selected that are resistant to 5-FOA (Dirnberger *et al.* 2008). Li and Fields confirmed the interaction of the p53 gene and T-antigen in the Yeast Two Hybrid system (Li and Fields 1993). The interaction partners are therefore very well suited as a positive control. In the case of the negative control, there is the possibility to transform empty vectors into the yeast and spot it on selective plates.

The advantage of the split-ubiquitin system is, that membrane proteins can be studied with the split ubiquitin system. So, these are not localized in the core (Stagljar *et al.* 1998). Another advantage is the two ubiquitin domains that are fused to a protein. This reduces steric hindrance. In the split ubiquitin system, the interaction is detected by the ubiquitin-specific proteases, which separate the reporter gene from the C-terminal domain. In the classic Yeast Two Hybrid System, the interaction is detected only by transcription initiation (Stagljar *et al.* 1998).

4. Discussion

In this thesis, the transformation of bait and prey constructs into yeast was made possible. It could be shown that the TEF1 promoter is very strong and prone to overexpression. Unfortunately, no phenotype could be determined in the interaction, as nothing could be done on the plates except of the TEF1 promoter. Due to lack of time, the test could not be further investigated and analyzed. However, the Split- Ubiquitin system is a very useful technique to identify protein-protein interactions.

5. References

Agola, J. O.; Jim, P. A.; Ward, H. H.; Basuray, S.; Wandinger-Ness, A. (2011): Rab GTPases as regulators of endocytosis, targets of disease and therapeutic opportunities. In *Clinical genetics* 80 (4), pp. 305–318. DOI: 10.1111/j.1399-0004.2011.01724.x.

Ao, X.; Zou, L.; Wu, Y. (2014): Regulation of autophagy by the Rab GTPase network. In *Cell death and differentiation* 21 (3), pp. 348–358. DOI: 10.1038/cdd.2013.187.

Awan, M. Umer Farooq; Deng, Yulin (2014): Role of autophagy and its significance in cellular homeostasis. In *Applied microbiology and biotechnology* 98 (12), pp. 5319–5328. DOI: 10.1007/s00253-014-5721-8.

Bachmair, A.; Finley, D.; Varshavsky, A. (1986): In vivo half-life of a protein is a function of its amino-terminal residue. In *Science (New York, N.Y.)* 234 (4773), pp. 179–186.

Bartel, Paul L.; Fields, Stanley (Eds.) (1997): The yeast two-hybrid system. New York: Oxford Univ. Press (Advances in molecular biology).

Blazeck, John; Garg, Rishi; Reed, Ben; Alper, Hal S. (2012): Controlling promoter strength and regulation in Saccharomyces cerevisiae using synthetic hybrid promoters. In *Biotechnology and bioengineering* 109 (11), pp. 2884–2895. DOI: 10.1002/bit.24552.

Brent, R.; Ptashne, M. (1985): A eukaryotic transcriptional activator bearing the DNA specificity of a prokaryotic repressor. In *Cell* 43 (3 Pt 2), pp. 729–736.

Brückner, Anna; Polge, Cécile; Lentze, Nicolas; Auerbach, Daniel; Schlattner, Uwe (2009): Yeast two-hybrid, a powerful tool for systems biology. In *International journal of molecular sciences* 10 (6), pp. 2763–2788. DOI: 10.3390/ijms10062763.

Dell'Angelica, E. C.; Mullins, C.; Caplan, S.; Bonifacino, J. S. (2000): Lysosome-related organelles. In *FASEB journal : official publication of the Federation of American Societies for Experimental Biology* 14 (10), pp. 1265–1278.

Dikic, Ivan; Elazar, Zvulun (2018): Mechanism and medical implications of mammalian autophagy. In *Nature reviews. Molecular cell biology* 19 (6), pp. 349–364. DOI: 10.1038/s41580-018-0003-4.

Dirnberger, Dietmar; Messerschmid, Monika; Baumeister, Ralf (2008): An optimized split-ubiquitin cDNA-library screening system to identify novel interactors of the human Frizzled 1 receptor. In *Nucleic acids research* 36 (6), e37. DOI: 10.1093/nar/gkm1163.

5. References

Eskelinen, Eeva-Liisa; Saftig, Paul (2009): Autophagy: a lysosomal degradation pathway with a central role in health and disease. In *Biochimica et biophysica acta* 1793 (4), pp. 664–673. DOI: 10.1016/j.bbamcr.2008.07.014.

Fang, Fang; Salmon, Kirsty; Shen, Michael W. Y.; Aeling, Kimberly A.; Ito, Elaine; Irwin, Becky et al. (2011): A vector set for systematic metabolic engineering in Saccharomyces cerevisiae. In *Yeast (Chichester, England)* 28 (2), pp. 123–136. DOI: 10.1002/yea.1824.

Fields, S.; Song, O. (1989): A novel genetic system to detect protein-protein interactions. In *Nature* 340 (6230), pp. 245–246. DOI: 10.1038/340245a0.

Golemis, E. A.; Serebriiskii, I.; Finley, R. L.; Kolonin, M. G.; Gyuris, J.; Brent, R. (2001): Interaction trap/two-hybrid system to identify interacting proteins. In *Current protocols in cell biology* Chapter 17, Unit 17.3. DOI: 10.1002/0471143030.cb1703s08.

Hermann, Greg J.; Schroeder, Lena K.; Hieb, Caroline A.; Kershner, Aaron M.; Rabbitts, Beverley M.; Fonarev, Paul et al. (2005): Genetic analysis of lysosomal trafficking in Caenorhabditis elegans. In *Molecular biology of the cell* 16 (7), pp. 3273–3288. DOI: 10.1091/mbc.e05-01-0060.

Hershko, Avram (2005): The ubiquitin system for protein degradation and some of its roles in the control of the cell-division cycle (Nobel lecture). In *Angewandte Chemie (International ed. in English)* 44 (37), pp. 5932–5943. DOI: 10.1002/anie.200501724.

Hirst, M.; Ho, C.; Sabourin, L.; Rudnicki, M.; Penn, L.; Sadowski, I. (2001): A two-hybrid system for transactivator bait proteins. In *Proceedings of the National Academy of Sciences of the United States of America* 98 (15), pp. 8726–8731. DOI: 10.1073/pnas.141413598.

Johnsson, N.; Varshavsky, A. (1994): Split ubiquitin as a sensor of protein interactions in vivo. In *Proceedings of the National Academy of Sciences* 91 (22), pp. 10340–10344.

Kitamoto, N.; Matsui, J.; Kawai, Y.; Kato, A.; Yoshino, S.; Ohmiya, K.; Tsukagoshi, N. (1998): Utilization of the TEF1-a gene (TEF1) promoter for expression of polygalacturonase genes, pgaA and pgaB, in Aspergillus oryzae. In *Applied microbiology and biotechnology* 50 (1), pp. 85–92. DOI: 10.1007/s002530051260.

Li, Qiu-Ping; Wang, Shuai; Gou, Jin-Ying (2017): A split ubiquitin system to reveal topology and released peptides of membrane proteins. In *BMC biotechnology* 17 (1), p. 69. DOI: 10.1186/s12896-017-0391-0.

Mumberg, D.; Müller, R.; Funk, M. (1995): Yeast vectors for the controlled expression of heterologous proteins in different genetic backgrounds. In *Gene* 156 (1), pp. 119–122.

5. References

Peng, Bingyin; Williams, Thomas C.; Henry, Matthew; Nielsen, Lars K.; Vickers, Claudia E. (2015): Controlling heterologous gene expression in yeast cell factories on different carbon substrates and across the diauxic shift: a comparison of yeast promoter activities. In *Microbial cell factories* 14, p. 91. DOI: 10.1186/s12934-015-0278-5.

Popovic, Doris; Akutsu, Masato; Novak, Ivana; Harper, J. Wade; Behrends, Christian; Dikic, Ivan (2012): Rab GTPase-activating proteins in autophagy: regulation of endocytic and autophagy pathways by direct binding to human ATG8 modifiers. In *Molecular and cellular biology* 32 (9), pp. 1733–1744. DOI: 10.1128/MCB.06717-11.

Qi, Hua; Xia, Fan-Nv; Xie, Li-Juan; Yu, Lu-Jun; Chen, Qin-Fang; Zhuang, Xiao-Hong et al. (2017): TRAF Family Proteins Regulate Autophagy Dynamics by Modulating AUTOPHAGY PROTEIN6 Stability in Arabidopsis. In *The Plant cell* 29 (4), pp. 890–911. DOI: 10.1105/tpc.17.00056.

Qin, Xiaoyu; Wang, Jiongyi; Wang, Xinxin; Liu, Feng; Jiang, Bin; Zhang, Yanjie (2017): Targeting Rabs as a novel therapeutic strategy for cancer therapy. In *Drug discovery today* 22 (8), pp. 1139–1147. DOI: 10.1016/j.drudis.2017.03.012.

Stagljar, I.; Korostensky, C.; Johnsson, N.; te Heesen, S. (1998): A genetic system based on split-ubiquitin for the analysis of interactions between membrane proteins in vivo. In *Proceedings of the National Academy of Sciences* 95 (9), pp. 5187–5192.

Stenmark, Harald; Olkkonen, Vesa M. (2001): The Rab GTPase family. In *Genome Biology* 2 (5), reviews3007.1-7.

Tang, Bor Luen (2017): Rabs, Membrane Dynamics, and Parkinson's Disease. In *Journal of cellular physiology* 232 (7), pp. 1626–1633. DOI: 10.1002/jcp.25713.

Wellhausen, Anne; Lehming, Norbert (1999): Analysis of the in vivo interaction between a basic repressor and an acidic activator. In *FEBS Letters* 453 (3), pp. 299–304. DOI: 10.1016/S0014-5793(99)00718-8.

Zenker, Martin; Mayerle, Julia; Lerch, Markus M.; Tagariello, Andreas; Zerres, Klaus; Durie, Peter R. et al. (2005): Deficiency of UBR1, a ubiquitin ligase of the N-end rule pathway, causes pancreatic dysfunction, malformations and mental retardation (Johanson-Blizzard syndrome). In *Nature genetics* 37 (12), pp. 1345–1350. DOI: 10.1038/ng1681.

6. Appendix

YAPD Medium

10 g Bacto-yeast extract

20 g Bacto-peptone

20 g Glucose

40 mg Adenine sulfate

water qsp 1 liter

adjust pH to 6,0 with HCl

autoclave

SC Medium

<u>1 l bottle:</u>

20 g Agar Bacteriological, water qsp - 400 ml

autoclave

<u>500 ml bottle:</u>

7 g nitrogenbase

1,4 g amino acid powder (-Leu -Trp -His -Ura)

qsp 500 ml H_2O, adjust pH to 5,9 with NaOH and autoclave.

Transfer Aminoacid mix to agar bottle and cool to 50 °C.

Add 100 ml 20 % glucose (s.f.) and 10 ml of 200 mg/100 ml Uracil.

Dry at least 2 days.

6. Appendix

Plates containing 3-Aminotriazole (3-AT)

Follow the recipe for SC Medium from above supplemented with the appropriate amino acids.

Cool to approximately 50-60 °C and add 3-AT as powders.

Stir to dissolve (few minutes), then pour plates without further adjusting the pH.

3-AT #09540 Fluka   84,08 g/mol

0,5 mM plates	add	42 mg 3-AT in 1 Liter SC Medium
5 mM plates	add	420 mg 3-AT in 1 Liter SC Medium
10 mM plates	add	840 mg 3-AT in 1 Liter SC Medium
25 mM plates	add	2,1 g 3-AT in 1 Liter SC Medium
50 mM plates	add	4,2 g 3-AT in 1 Liter SC Medium
100 mM plates	add	8,4 g 3-AT in 1 Liter SC Medium

Luria-Bertani (LB) Medium

5 g Tryptone

2,5 g Yeast extract

5 g NaCl

Filled up with 500 ml distilled water

For plates: ad 18 g/l Agar

Add MQ - H_2O and autoclave

6. Appendix

Oligonucleotides

Tab. 6: List of used Oligonucleotides.

Primers	Sequence	Description
oGUF49	GAGGCGCGCCAGAGGAGTACA CACGGGAC	forward primer for yeast SSB1 promoter to amplify a 500 bp fragment introducing an AscI/MluI site upstream
oGUF50	GAGGTACCGGATCCTTTGTT CAATTAAAATACTGTAATGAT	reverse primer for yeast SSB1 promoter to amplify a 500 bp fragment introducing a BamHI and KpnI site downstream
oGUF51	GAGGCGCGCCATAGTGGGTAG ATAACCTTAAACTATTTATTTA GAG	forward primer for yeast SSA1 promoter to amplify a 566 bp fragment introducing an AscI/MluI site upstream
oGUF52	GAGGTACCGGATCCATTATCTG TTATTTACTTGAATTTTTGTTTC TTG	reverse primer for yeast SSA1 promoter to amplify a 566 bp fragment introducing a BamHI and KpnI site downstream
oGUF53	GATCGCGGCCGCTGCATTTCTT CCAAGTCATCTGCTC	reverse primer for C35B1.2a binds at Aa301, 58°C, without stop and introduces a NotI site
oGQ1232	GAAGTGTCAACAACGTATCTAC CAACGA	Reverse Primer to sequence pFR5012_pDB Trp TEF1 GFP NB Cub His3x
oGQ1514	CGTCTTCTAGCTGCTTACCGGC AAAGATC	Reverse Primer to sequence pDB Trp – SSA1 – C35B1.2a and pDB Trp – SSB1 – C35B1.2a
oGQ3032	GCAAGCTTGGTACCGGATCCAT GGCC	Forward Primer to sequence pFR5012_pDB Trp TEF1 GFP NB Cub His3x
oGQ3078	CGTTCCCTTTCTTCCTTGTTTC	Forward primer to sequence pNub frame 2 – glo-1(QL)
oGQ3250	GGGGGACCGGTAGACGCCTCC GAGATCCCAAA	Forward Primer to sequence pDB Trp – TEF1 – C35B1.2a
oGQ3407	CTCGAGAATTCTGGCCAAGGAA AGCGGCGTA	Forward Primer to sequence pFR5012_pDB Trp TEF1 GFP NB Cub His3x
oGQ3572	TTCATTTTTCTTGTTCTATTA	Forward Primer to sequence pFR5012_pDB Trp TEF1 GFP NB Cub His3x
oGQ3573	CCTTCTGAATGTTGTAATCAG	Reverse Primer to sequence pDB Trp – SSA1 – C35B1.2a (Aa 95-301) and pDB Trp – SSB1 – C35B1.2a (Aa 95-301)
oGQ4067	AACTCGACCATTTCGACATTGG A	Forward Primer to sequence pDB Trp – SSA1 – C35B1.2a and pDB Trp – SSB1 – C35B1.2a

6. Appendix

Plasmids

Tab. 7: List of plasmids used in this Thesis.

Backbone	Enzymes	Insert	Bacterial selection	Yeast selection	Result	Source
pNub frame 2	NcoI/NotI	glo-1(QL)	Ampicillin	LEU2	pNub frame 2 – glo1(QL)	S.König, personal comment
pGEMT	Ligation	C35B1.2a (Aa 95-301)	Ampicillin	-	pGEMT - C35B1.2a (Aa 95-301)	Promega
pGEMT	Ligation	SSA1	Ampicillin	-	pGEMT - SSA1	Promega
pGEMT	Ligation	SSB1	Ampicillin	-	pGEMT - SSB1	Promega
pDB Trp - TEF1 – GFP NB - Cub His 3*	Acc65I/NotI	C35B1.2a	Kanamycine	TRP1	pDB Trp - TEF1 – C35B1.2a Cub His 3*	T. Blume
pDB Trp - TEF1 – GFP NB - Cub His 3*	NcoI/NotI	C35B1.2a (Aa 95-301)	Kanamycine	TRP1	pDB Trp - TEF1 – C35B1.2a (Aa 95-301) Cub His 3*	T.Blume
pDB Trp - TEF1 – C35B1.2a Cub His 3*	Acc65I/NheI	SSA1	Kanamycine	TRP1	pDB Trp – SSA1 – C35B1.2a Cub His 3*	This Thesis
pDB Trp - TEF1 – C35B1.2a Cub His 3*	Acc65I/NheI	SSB1	Kanamycine	TRP1	pDB Trp – SSB1 – C35B1.2a Cub His 3*	This Thesis
pDB Trp - TEF1 – C35B1.2a (Aa 95-301) Cub His 3*	MluI/Acc65I	SSA1	Kanamycine	TRP1	pDB Trp – SSA1 – C35B1.2a (Aa 95-301) Cub His 3*	This Thesis
pDB Trp - TEF1 – C35B1.2a (Aa 95-301) Cub His 3*	MluI/Acc65I	SSB1	Kanamycine	TRP1	pDB Trp – SSB1 – C35B1.2a (Aa 95-301) Cub His 3*	This Thesis